Learning to Handbook:

C000042568

Detailed Step By Step Guide on How to Sail; The Salient Rules (Dos & Don'ts); The Required Preparation Plus Some Vital Questions You Would Want to Ask & Lots More

By

Wanda K. Romani

Copyright@2020

TABLE OF CONTENTS

CHAPTER ONE

The Vital Reasons You Should
Learn To Sail

I experienced childhood in San
Diego, and I feel about cruising the
manner in which surfers talk about
surfing. Cruising is particularly
energizing for first--timers.
Probably the coolest thing is
cruising upwind, even just around
the cove. You're going to feel the
breeze all over, you will hear the
sail. A major pontoon drops by,
and there's a wake. You hear the

waves sprinkling and feel the vessel heel aside. Out of nowhere two or three dolphins jump out of the water. And so much stuff is tangible data that encourages you figure out what to do. The sound of the sail lets you know whether it's luffing, if it's not cut appropriately. The manner in which the water moves predicts a coming movement of the breeze. That is what's exceptional about cruising: The association you make with nature is an outcome of your creation the pontoon move. Once in a while, when you have everything great, it goes calm, and everything you can hear is the frame experiencing the water and the breeze pushing past you. It's basic. Otherworldly, even.

Acing cruising fundamentals implies being acquainted with basic option to precede cruising rules. One thing to recall is that there are distinctive options to precede cruising guidelines for boats instead of powerboats.

CHAPTER TWO

ESSENTIAL FACTS ABOUT SAILING YOU SHOULD KNOW

Discover a Place to Learn

Between the American Sailing
Association and U.S. Cruising, the
two primary ensuring bodies, there
are a couple hundred cruising
schools around the nation—even
in unforeseen spots like Oklahoma
and Arizona. Apprentice's courses
frequently most recent two days, so
you can discover end of the week
bundles. Hope to pay about $500
per individual. At more significant
levels U.S. Cruising stresses
hustling and the ASA joy sailing,

yet basic classes are pretty much the equivalent.

Comprehend What's Happening

Cruising is rich with language, custom, and legend. You'll better retain the complexities in the event that you know the nuts and bolts early.

How Wind Moves a Boat

A sail loaded with wind shapes an airfoil and pushes the vessel with lift, the manner in which a plane's wing does (aside from across water, instead of into the air). Crafted by cruising is to position, or trim, the sails to augment lift toward the path you need to go.

When you've raised the sails utilizing the lines—never state "ropes"— called halyards (A), they're cut utilizing the sheets (D), which rotate the blast (E) between the port and starboard—that is, left and right—sides of the pontoon. On a two-sail vessel like this one (called a sloop), the accentuation is on the mainsail (B), the sail closer the harsh (H), which is the rearmost or toward the back piece of the pontoon. The littler jib (C), closer the bow (F), at the front, likewise rotates—yet as another mariner you'll be centered around the mainsail.

The essential thought: You utilize the turner (G) to move the rudder (I) and edge the pontoon so it is opposite to the breeze. Utilize the sheets to edge the mainsail so it loads up with wind. In the bowing airfoil shape, air moving over the more extended, bended side moves quicker than air streaming by the opposite side, creating lift.

CHAPTER THREE

SAILING RULES (DOS & DON'TS) FOR YOU

1. Continuously keep up an appropriate post by sight just as hearing to abstain from slamming into different pontoons

2. Keep up a sheltered speed consistently with the goal that you stay in charge of your vessel

3. Utilize sound judgment when surveying danger of impact with different vessels close and around you

4. Port attach offers approach to starboard tack: If two boats are moving toward one another and

the breeze is on an alternate side of each pontoon, at that point cruising decides are that the boat which has the breeze on the port side should consistently give option to proceed to the next. (The port side is the left-hand side of the pontoon when you are confronting the front.)

5. Windward offers approach to leeward: If two boats are moving toward one another and the breeze is on a similar side of each pontoon, at that point cruising

decides are that the vessel which is to windward (the course of the breeze) must give the option to proceed to the vessel which is leeward (the other way of the breeze).

6. In the event that you are in danger of crashing into another vessel and all else falls flat, at that point concurred cruising decides are that whichever pontoon has the other vessel on its starboard side must yield option to proceed. (The starboard is the right-hand side of

the vessel when you are
confronting the front.)

7. Any vessel overwhelming
another ought to consistently keep
off the beaten path of the vessel
being surpassed.

8. A boat ought to consistently
keep off the beaten path of any
vessel that is: a) not under order,
b) confined in its capacity to move,
and c) occupied with fishing

9. When going through a restricted channel, cruising guidelines are to keep as near the external edge as could reasonably be expected.

10. Non-business powerboats normally offer approach to boats, except if the boat is overwhelming it. In any case, general cruising guidelines are additionally that boats should attempt to avoid the method of enormous vessels and ferryboats that may think that its

harder to slow or alter course—

particularly in restricted channels

CHAPTER FOUR

THE KEY/ESSENTIAL STEPS OF SAILING

The Key/Essential Steps of Sailing

Cruising includes both explicit information and aptitudes. Coming up next are the fundamental strides of figuring out how to cruise as much as possible learn while not really on a vessel. You don't need to follow this request; avoid ahead on the off chance that you definitely know a portion of the fundamentals. In case you're for the most part new to cruising, the following will be useful to you right away:

1. Understand Basic Sailing Terms.

To get into cruising, you need to comprehend the words that are utilized to discuss the boat and the abilities used to cruise. Start here with a survey of essential cruising terms. Try not to stress over retaining everything the same number of these terms and ideas will become more clear as you read on about how to do it.

2. Learn the Parts of the Boat.

Before you go on the pontoon, it's useful to realize the words utilized in various pieces of the vessel.

Regardless of whether you have a teacher, the person in question won't state "Get that rope over yonder and pull it," however rather will say "Take in the jib sheet!"

3. Start an Online Course.

Presently you're prepared to get familiar with what every one of those pieces of the pontoon are utilized for. Here you can begin an online figure out how to-cruise course by becoming familiar with the pieces of the pontoon alongside a great deal of photographs, so you'll see what to do.

4.Rig the Boat.

Peruse to go cruising now? Hold it a moment you need to fix the vessel first by putting on sails and making different arrangements.

Alright, presently you have the vessel prepared so what do you do now to cause it to go? Deal with the sails to go toward the path you need by learning essential cruising procedures.

5. Discover How to Manoeuvre.

Cruising a set way is sensibly simple, yet in the long run, you'll need to alter course. That frequently includes attaching and gybing. Pause for a minute to realize what's associated with these basic moves.

6. Recover From a Capsize.

Presently you have the nuts and bolts down. In any case, did anybody ever disclose to you that little boats frequently spill if the breeze is blasting? Be readied and cautiously perceive how to recuperate from an upset.

7. Dock or Anchor the Boat.

Presently you're out there cruising and you have the vessel levelled out. Figure out how to speed up, moor or stay the pontoon and utilize a portion of the hardware you've disregarded up until this point. Investigate a portion of these extra cruising aptitudes.

8. Practice Tying Knots.

For a huge number of years, mariners have utilized occasions where it is cold or coming down by doing things like tying ties. Bunches are significant on a boat and you should learn probably some fundamental cruising bunches to cruise by any stretch of the imagination.

9. Sail Safely.

Now, in addition to rehearse on the water, you're all set. Be that as it may, it's acceptable to recall that water is a hazardous spot. Gain proficiency with the nuts and bolts about cruising security. Remaining safe makes it simpler to continue having a ton of fun out there.

CHAPTER FIVE

SOME VITAL AS WELL AS GOOD QUESTIONS BEGINNER SAILOR WOULD WANT TO ASK RIGHT AWAY

Inquiries From the First-Time Sailor

Imagine a scenario in which there's a whirlwind. Will the vessel overturn?

When there's a major blast, you basically transform into the breeze, which stops the vessel.
Furthermore, fortunately, the overwhelming counterweight under a keelboat's structure makes them entirely difficult to overturn. So while a solid blast may make the vessel heel, or lean, you'll be fine.

Will I get nauseous?

You may—however while -learning, you'll be near shore and having the option to see land by and large makes a difference. In case you're going to upchuck, to assist all included, head to the -leeward side of the pontoon.

On the off chance that I fall over the edge, can the vessel make it back to me so as to spare me?

Make it realized that you've hit the water and somebody will throw you a buoyancy gadget (in the event that you aren't as of now wearing one). At that point, it's going to quickly seem as though the vessel is cruising ceaselessly. Try not to go ballistic. It just requires some investment to turn a boat around.

How hard is it to function the sails?

Wind-filled sails can take genuine solidarity to oversee, however devices like pulleys and winches give mariners a mechanical preferred position. As an easygoing interest, anybody can cruise. (Dashing is another issue.)

Would it be advisable for you to Buy a Boat?

As a matter of fact, yes. In the event that you've gotten the bug, a decent section highlight vessel possession is a Sunfish, a 14-foot dinghy with a solitary sail. New, a Sunfish can be had for about $4,500, however they've been around since the mid 1950s—with a few hundred thousand in presence, there's in any event a couple out there fit as a fiddle at a much lower cost.

CHAPTER SIX

THE REQUIRED PREPARATION PLUS THE TYING OFF THE VERY LINE

Getting Prepared

Any captain taking you out for an exercise will have security gear and route equipment. Your principle work is to dress for the event. So: Picture a mariner in your brain—at that point don't dress that way. You needn't bother with a link weave sweater and channel. Check the figure and wear the layers you'd wear ashore—in addition to make a point to have these four things.

1. Shoes: Two prerequisites for deck shoes: They should be grippy and non-checking. With a Vibram Wavegrip sole, the Columbia Men's Force 12 PFGs ($130) meet both and offer a third favourable position: They look incredible.

2. Shades: Polarized shades cut glare so you can perceive how the water is moving, helping you read the breeze. Costa's optics are best in class. We suggest its Slack Tide conceals ($259) with reflect focal points, which increment differentiate between inconspicuous changes in water conditions.

3. Gloves: Expensive cruising gloves can push $50. Showa Atlas 300 substantial gloves are about $21 on Amazon—for a 12-pack. They work similarly too, and in the event that you need your fingertips uncovered (to improve mastery), you won't feel awful taking a -scissor to them.

4. Coat: A layer that can rise up to breeze is non--negotiable. Helly Hansen's HP Fjord ($200) is a fantastic obstruction to both breeze and water, and there's simply enough heave to it to offer a little warmth, as well.

5. The Logbook: The logbook is the place you monitor the classes you've finished and the hours you spend on the water. On the off chance that you ever need to contract a pontoon or become an educator, it's a convenient method

to show understanding.
Furthermore, regardless of whether
you don't, keeping an
unmistakable record of
achievements is a misjudged joy.

Tying Off the Very Line

Sooner or later you'll be
approached to tie down a line to a
fitting. Here's the manner by
which, utilizing a basic bunch
called a projection hitch.

1. Beginning with the horn farthest
from the heap, fold the line over

the two horns. (Just a single time—
more expands the chances of
sticking.)

2. Make in any event two figure-
eight pivots the fitting.

3. Secure the free stopping point by
tucking it under the last turn.

THE END

Printed in Great Britain
by Amazon

22823364R10030